21st Century Junior Library

UNDERSTANDING EARTHQUAKES AND TSUNAMIS

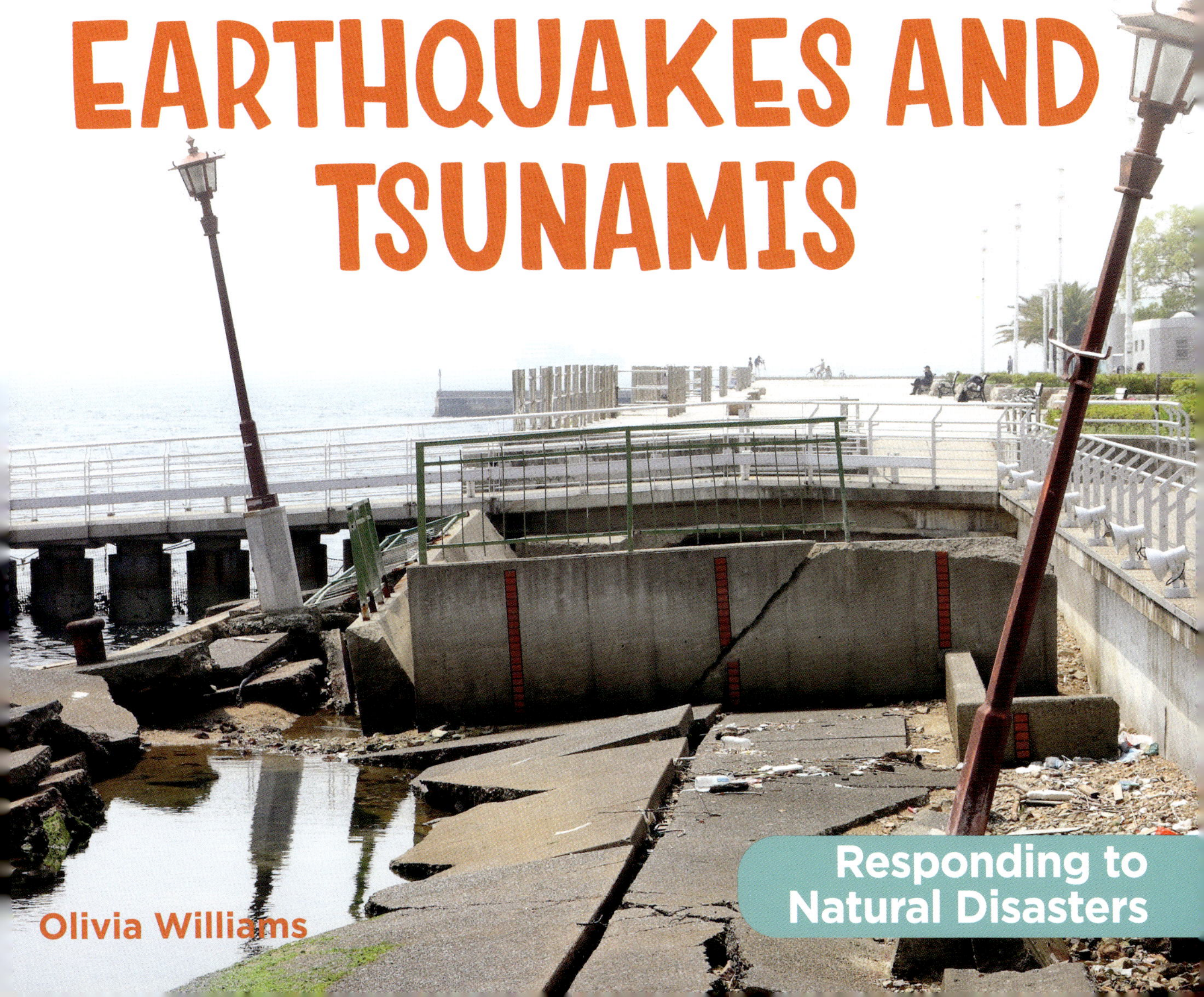

Responding to Natural Disasters

Olivia Williams

Published in the United States of America by:

CHERRY LAKE PRESS
2395 South Huron Parkway, Suite 200, Ann Arbor, Michigan 48104
www.cherrylakepress.com

Reading Adviser: Beth Walker Gambro, MS, Ed., Reading Consultant, Yorkville, IL
Photo Credits: © Tupungato/Shutterstock.com, cover, 1; © Seba Rem/Shutterstock.com, 5; © VicPhotoria/Shutterstock.com, 6; © Dudarev Mikhail/Shutterstock.com, 7; © Paolo Sartorio/Shutterstock.com, 9; © Keystone Press/Alamy Stock Photo; © Fly_and_Dive/Shutterstock.com, 12, 13; © VicPhotoria/Shutterstock.com, 14 [left, right]; © Denis Belitsky/Shutterstock.com, 15 [top]; © ianmitchinson/Shutterstock.com, 15 [bottom, left, right]; © Michael Barajas/Shutterstock.com, 16; © simone tognon/Shutterstock.com, 17 [top]; © Youkonton/Shutterstock.com, 17 [bottom]; © Tomas Nevesely/Shutterstock.com, 18

Cherry Lake Press is an imprint of Cherry Lake Publishing Group.

Library of Congress Cataloging-in-Publication Data
Names: Williams, Olivia (Writer of children's books), author.
Title: Understanding earthquakes and tsunamis / Olivia Williams.
Description: Ann Arbor, MI : Cherry Lake Publishing, [2022] | Series: Responding to natural disasters | Audience: Grades 2-3
Identifiers: LCCN 2021043073 | ISBN 9781534199156 (hardcover) | ISBN 9781668900291 (paperback) | ISBN 9781668901731 (pdf) | ISBN 9781668906057 (ebook)
Subjects: LCSH: Earthquakes—Juvenile literature. | Tsunamis—Juvenile literature.
Classification: LCC QE521.3 .W55 2022 | DDC 363.34/95—dc23/eng/20211005
LC record available at https://lccn.loc.gov/2021043073

Cherry Lake Press would like to acknowledge the work of the Partnership for 21st Century Learning, a Network of Battelle for Kids. Please visit http://www.battelleforkids.org/networks/p21 for more information.

Printed in the United States of America
Corporate Graphics

ABOUT THE AUTHOR

Dr. Olivia Williams is from mid-Wales in the United Kingdom. She has traveled to many countries to help people after natural disasters. She has felt earthquakes and aftershocks, and seen damage from floods, typhoons, and wildfires. The natural environment can be harsh but also beautiful. Olivia loves to hike and run in the mountains and photograph the incredible things she sees.

CHERRY LAKE PRESS

CONTENTS

WHAT ARE EARTHQUAKES AND TSUNAMIS?

Earth is made up of four layers. These layers are the inner core, outer core, mantle, and crust. These layers move in ways that we cannot always see. The crust is the layer that we stand on.

The crust is made of **tectonic plates**. Tectonic plates are huge moving rocks. These plates are about 100 miles (160 kilometers) below our feet. Scientists think that tectonic plates are about

Scientists have been able to study only one of the four layers of Earth in person.

4.6 billion years old. That means they are older than all of us. They are even older than dinosaurs!

The plates move less than 1 inch (2.5 centimeters) each year. We barely notice this slow movement. In fact, they move as slowly as our fingernails grow! When tectonic

Tectonic plates are like puzzle pieces that move and bump into each other.

The movements and bumps of tectonic plates are what cause earthquakes and tsunamis.

plates crash together, they cause an earthquake. An earthquake is a type of **natural disaster**. Sometimes earthquakes happen underwater. When an earthquake happens on the ocean floor, it can cause a group of large waves called a **tsunami**. Sometimes tectonic plates crash so hard that they are forced upward. This is how mountains and islands form. Scientists call the place where tectonic plates crash the **fault line**.

Only some places experience strong earthquakes. California is famous because it has the San Andreas Fault. Houses and buildings are built on top of it. People's homes could be destroyed if there were an earthquake at this fault line. Alaska has strong earthquakes too. Alaskan earthquakes happen more often than those in California.

Make a Guess!

Imagine you're watching television. A news report says that an earthquake is likely to happen where you live. What do you think you and your family could do to prepare for the earthquake and stay safe?

During an earthquake, stay away from sharp and breakable objects, such as windows or kitchen utensils.

Only 15 minutes after the devastating 1960 Chile earthquake ended, a massive 80-foot-high (24 meters) tsunami headed toward the country's coast.

Earthquakes are measured by their strength. The measurement tools scientists use are called **seismographs**. Seismographs give earthquakes a number that tells how strong they are. We probably wouldn't feel a level 3 earthquake, but we definitely would feel a level 8 earthquake. It would destroy many of the things around us. In 1960, the strongest earthquake ever recorded hit Chile in South America. The Chile earthquake was a level 9.5, and it lasted 10 minutes!

HOW EARTHQUAKES AND TSUNAMIS AFFECT THE PLANET AND PEOPLE

We know an earthquake is happening when we feel the ground shake. Earthquakes can be very scary and destructive. Sometimes things fall from shelves or buildings collapse. Often, roads crack and create big holes that cars can fall into. Earthquakes can even lead to landslides and fires.

大型
(マイクロを除く)

Strong earthquakes can cause a tsunami. A tsunami can travel up to 500 miles (800 km) per hour. This is as quick as the world's fastest car!

Ask Questions!

UNICEF is a humanitarian organization that helps and protects children after a natural disaster. Today, it stands for the United Nations Children's Fund. The organization provides access to clean water, education, food, and more. If you had the opportunity to meet a UNICEF partner, what would you ask them?

The word *tsunami* comes from two Japanese words: *tsu*, which means "harbor," and *nami*, which means "wave."

Tsunamis can also be 100 feet (30 m) high. That's the same height as five and a half giraffes stacked on top of one another!

Think!

Think about the vulnerable people you see in your neighborhood. Can you suggest the ways that you and your family could help them after an earthquake?

To compare, a typical wave at the beach reaches about 3 feet (1 m) high.

In 2004, a strong 9.1 earthquake occurred under the Indian Ocean. The earthquake created a tsunami. The tsunami killed thousands of people in Thailand. Homes and other buildings were destroyed. People lost many of the things that they needed to survive.

If you suspect a tsunami, evacuate immediately. Try to get to higher ground and as far away from the coast as possible.

PROTECTING AGAINST AND RECOVERING FROM EARTHQUAKES AND TSUNAMIS

Scientists studying earthquakes and fault lines are called **seismologists**. Seismologists try to figure out when an earthquake might happen. Knowing when an earthquake might hit can keep people safe. Sometimes seismologists create maps of earthquake areas. Rescue teams use these maps to help people and animals trapped after an earthquake.

After an earthquake or tsunami, people from all over the world travel to help. They are called humanitarians or **emergency response teams**. In 2015, a big 7.8 earthquake hit Nepal and killed thousands of people. About 70 countries from around the world sent teams to help. Nepalis also worked very hard to help their people.

People learn from natural disasters. Nepalis are now rebuilding their homes to be more **flexible**. Flexible buildings sway during an earthquake and don't collapse. These buildings help keep people safe.

Create!

Children were upset and afraid after the earthquake in Nepal. Emergency response teams helped the children by encouraging them to create art. Why do you think art can help people to express their feelings?

ACTIVITY

Making a Plan

Communication and planning are important factors when faced with a natural disaster. Sit down with your family and establish a plan. Read through the list below to help you get started!

- Research common natural disasters that may affect your area.
- Establish a meeting place in case you get separated and need to reconnect.
- Assign roles to each person (e.g., Who is in charge of grabbing go-bags? Who is in charge of the pets?).
- Write down important contact information and make several copies.
- Draw exit plans for each room in your house.

GLOSSARY

emergency response teams (ih-MUHR-juhn-see rih-SPAHNSS TEEMZ) groups of people with special skills, such as doctors, rescuers, and pilots, who help save people hurt by natural disasters

fault line (FAHLT LYNE) a long crack in the surface of Earth; earthquakes usually occur along fault lines

flexible (FLEK-suh-buhl) bends easily without breaking

humanitarian (hyoo-MAH-nuh-tahr-ee-uhn) a person who tries to save lives after a natural disaster or a war

natural disaster (NAH-chuh-ruhl dih-ZAH-stuhr) natural event such as a flood, earthquake, or hurricane that causes great damage

seismographs (SYZE-muh-grafz) instruments that measure and record details of earthquakes, such as force and duration

seismologists (size-MAH-luh-jists) scientists who study natural events like earthquakes; they try to figure out when an earthquake will happen and how strong the earthquake will be

tectonic plates (tek-TAH-nik PLAYTS) enormous slabs of moving rock under the Earth's crust

tsunami (soo-NAH-mee) long and high wave in the ocean caused by an underwater earthquake

vulnerable (VUHL-nuh-ruh-buhl) capable of being easily hurt or injured, like small children, the elderly, or animals

FIND OUT MORE

Books

Drimmer, Stephanie Warren. *Ultimate Weatherpedia: The Most Complete Weather Reference Ever.* Washington, D.C.: National Geographic, 2019.

Everything Earthquakes and Tsunamis. Newark, DE: Baby Professor, 2020.

Websites

Earthquake—How Earthquakes Happen

https://youtu.be/ROVuhDENYh0
Watch and learn more about how earthquakes happen.

Plate Tectonics—Tectonic Plates Theory

https://youtu.be/MZ-sVQElqhE
Learn about the three different types of plate movements.

INDEX